# The Quest of the Leghorn
## A Poultry Breeder's Journey In Search of the Leghorn Chicken

*by Frederick H. Ayres*

**with an introduction by Jackson Chambers**

*This work contains material that was originally published in 1880.*

*This publication is within the Public Domain.*

*This edition is reprinted for educational purposes
and in accordance with all applicable Federal Laws.*

*Introduction Copyright 2017 by Jackson Chambers*

# Self Reliance Books

Get more historic titles on animal and stock breeding, gardening and old fashioned skills by visiting us at:

# http://selfreliancebooks.blogspot.com/

## *Introduction*

I am pleased to present yet another title in the "Chicken Breeds" series.

This volume is entitled "The Quest For The Leghorn".

In 1880, reknowned American poultry expert, Frederick H. Ayres, made a voyage to Italy for the express purpose of securing Leghorn Fowls and acquiring accurate knowledge of the breed in their native territory. This was probably the first voyage of considerable length ever made for the sole purpose of importing chickens. Mr. Ayres did not set out in "the Quest For The Leghorn" as a pleasure trip or for any other business. His sole purpose was to find the Leghorn breed in its home region and in its prosecution, he sailed over twelve thousand miles. This trip over the sea would have taken him over 100 days!

This is the story of Mr. Ayres' quest for the Leghorn chicken.

The work is in the Public Domain and is re-printed here in accordance with Federal Laws.

Though this work is a century old it contains much information on poultry that is still pertinent today.

As with all reprinted books of this age that are intended to perfectly reproduce the original edition, considerable pains and effort had to be undertaken to correct fading and sometimes outright damage to existing proofs of this title. At times, this task is quite monumental, requiring an almost total "rebuilding" of some pages from digital proofs of multiple copies. Despite this, imperfections still sometimes exist in the final proof and may detract from the visual appearance of the text.

I hope you enjoy reading this book as much as I enjoyed making it available to readers again.

Jackson Chambers

## PREFACE.

This little work makes no claim to special literary merit, nor to interest, save that which must attach to the mere recital of facts hitherto unknown to American fanciers. We have aimed throughout to give facts and let them stand without much help from preconceived theories. As we would write to a friend the account of what we learned, we have written for the inspection of the poultry fraternity.

# THE QUEST OF THE LEGHORN.

## How the Quest Came About.

The project of making a trip of over ten thousand miles, for the sole purpose of ascertaining the existence, or non-existence of fowls of a certain type of domestic poultry, is an entirely new departure in the annals of poultry culture. As such it seems worthy of a few words of introduction; a sort of reply to the "how came you so" of poultry breeders generally.

The first impetus in the general plan which finally was fulfilled in the voyage to Italy, came from the great poultry show which was for a few days a feature of the Paris Exposition of 1878. At that time Mr. F. H. Corbin, of Newington, was first becoming generally known through his "Improved" Plymouth Rock fowls; and as we knew that he made poultry breeding a business, the idea that he might join us, in the importation of stock selected at this exhibition, very naturally occurred. With this in view we made a trip to Newington and had a conference with Mr. Corbin. The plan, however failed to meet his approval and so, as there was no time for the correspondence incident to interesting a number of fanciers the project dropped.

For nearly a year other business claimed our attention, but with the summer of 1879 came a revival of the dormant idea, under a little different form. At this time our little pamphlet on "The Leghorn" had been followed by two books on "Brown Leghorns," by H. H. Stoddard and C. R. Harker, and the interest in this favorite variety had been still farther heightened. This seemed, then, the proper time for an effort for still greater improvement, and so the "Poultry World" was called into requisition to acquaint poultrymen with our design. This was in August, and in our first modest card we spoke of importing only Rose-Comb Leghorns. The inquiries for other varieties were, however, so numerous that when the September "World" came out we changed the announcement to the effect that orders for any known variety of Leghorn fowl would be taken and filled, as nearly as circumstances would allow. Responses to this last card were very prompt in making their appearance, and so ere the month was out we were on the Atlantic, bound east five thousand miles in quest of the Leghorn.

## What People Thought About It.

It is interesting to note the opinions of the best informed breeders and we here give a sample from the American Poultry Yard of what was said at the time of our going. Mr. Stoddard, is we believe, more thoroughly posted on poultry matters generally, than almost any other man in America, and this editorial notice represents the opinions of the most advanced fanciers, prior to the trip. The article reads as follows:

## A Novel Enterprise.

"The value of importation lies, to a great extent, in a feature which has but little consideration with the majority of breeders—the introduction of new blood in no degree related to that which has been longest bred in this country. Were there no gain in any other way, could we procure no better stock than we already have, on this ground alone the extra cost of importation would be well repaid.

In view of this state of the case, it is strange that no one has before taken the necessary steps to import fowls in quantity from foreign sources. It seems to have been left for Mr. F. H. Ayres, of Mystic River, Conn., whose advertisement appears in the August *Poultry World*, to put the machinery in motion for doing this work thoroughly and in a business like way. Before giving notice of his plans, Mr. Ayres spent a considerable time in ascertaining from shipmasters and others who have visited the Mediterranean and Italian ports, the cost of the fowls in their native countries, the rates of freight and passage, and all other particulars of the task he had set before himself.

This done, he was enabled to announce his intention of bringing from Italy a large stock of Leghorn fowls of all varieties, rose and single-combed.

Mr. Ayres's plan—which now is assured—is to sail for Italy about the 10th of September, and after buying the stock which is subscribed for, return at the earliest possible moment. Mr. Ayres will accompany the fowls and give them his personal care throughout the trip, till they are placed in the hands of the express company.

The fact that Mr. Ayres, unlike Captain Corcoran, is a good sailor and never sea-sick, insures that care and attention for lack of which so many birds suffer on sea voyages.

Such a chance to import stock of this variety will probably not occur again in years; for though a non-fancier could be commissoned to buy fowls, only a well-posted breeder can purchase them to advantage.

On his return Mr. Ayres will publish a book under the title of 'The Quest of the Leghorn,' and detail all he saw to interest poultrymen; and we may count on a very readable and instructive work.

All who wish to obtain Rose-Comb, White or Brown Leghorns, or add fresh blood to their present stock, will do well to communicate with Mr. Ayres at Mystic River, Conn.

We have given only the bare outlines of this project, and all interested should write for further important particulars."

## Making Preparations.

Such was in brief, the history of this novel quest up to the point of setting sail, but before the events we have so hastily sketched could transpire, a number of minor details needed attention. Before even the first announcement of the plan could be made public, a comparison of the various ways of getting to and from Italy, a reasonably close estimate of the cost of fowls and their freight with due allowance for incidental expenses must be told off. While searching for information on these points we naturally kept a bright lookout for any poultry stories from shipmasters who had been in the Western ocean trade, and picked up a few items which we jotted down in our note book as follows:

## Notes By The Way.

Our experience in getting new and noteworthy facts on poultry matters generally began even before we left New York for Europe. The vessel on which we had taken passage for Marseilles was detained several days by bad weather and meantime we one evening met Captain Curry of the bark Lizzie Curry, which had two days before come into port. Naturally the conversation turned on the novel purpose of our trip and after some discussion the Captain spoke of the Mediterranean trade in fowls and eggs. "You probably have no idea," said he, "of the trade there is in these ports. Why at Vigo, on the west coast of Spain, the steamers stop once or twice a week for shipments of fowls and eggs. These days are known as 'steamer days,' and the peasants come from far and near to send their stock. You may be sure that the trade that calls for the stoppage of two ocean steamers weekly is no trifle."

The Captain went on to describe some magnificent lustrous black fowls that he had kept on board the vessel for months until finally they " caught cold, had swelled heads and died." Our readers will recognize the handsome Black Spanish, and the common enemy, roup.

This Captain Curry is the son of the master of the vessel in which we sailed. The senior captain has been for eighteen years a Mediterranean trader, and from him we also extracted some bits of information on poultry topics.

Some eight years since he bought at Trieste, in Austria, a cock in general shape "like the Shanghaes," and gave him to a brother captain living on Long Island, New York. For several years thereafter he saw the bird at the owner's place, and from year to year noted how the tone of the whole flock was being raised by this influence. In this case no special clean breeding was attempted but the potent power of utterly fresh blood left to do its will, with results that show fully the value of such an occasional change.

Some fowls live to be old stagers in maritime experience, as is seen by the fact that a hen bought for the table, but afterwards kept as a pet, lived a year and a half among all the vicissitudes of weather and the changes of climate from New York to the ports of Southern France, again and again. This bird knew as well as any sailor when a barrel of spray was coming on board, and would scuttle for shelter just in the nick of time. She had no regular nest but dropped her eggs at any convenient point where they could not roll away.

# From New York to Marseilles.

After some delay in New York we at last cast off our moorings and under steam passed down the East river and the harbor and by noon we found ourselves outside the harbor with a light south-westerly wind. This continued for four days, and then the breeze strengthened materially and we ran to the Azores, or Western Islands, about twenty-five hundred miles in fourteen days. Here we lay well-nigh becalmed for a week and did not finally reach the straits of Gibraltar till the thirtieth day out. From this point land was in sight all the way as we coasted along about fifteen miles from the Spanish coast till Spain touched France. Here under Cape Sebastian, with the monastery in which Queen Isabella took refuge in full sight, we lay for eleven days in a gale of wind. But all things come to an end, and so the evening of November 24th saw us at anchor in the harbor at Marseilles after an unusually long passage, due chiefly to heavy weather.

# The Marseilles Zoo.

On landing in Marseilles we first sought a hotel where English was understood and were fortunate enough to stumble directly upon the American consul, Mr. Gould, to whom we had letters from a mutual friend. Through his unwearied kindness we were enable to see much of the old city, which would escape the eye of a stranger, and in various ways pass pleasantly the three days which intervened between our arrival and the sailing of the steamer for Leghorn. We do not purpose to inflict on our readers any recapitulation of the standard recitals of the guide books and letter writing travelers, but limit our recital to the fowls seen at the Zoological Garden.

The garden lies back of the city on the crown of a high hill, and is accessible from two sides. At the main entrance we face the artificial falls supplied by jets from the mouths of gigantic bulls, carved in stone, which lie with raised heads in front of the stone building through which access is had to the main grounds. Passing through into the main garden which occupies the crest of the hill we found a number of smooth walks and rustic seats, and in the open spaces flower beds and shrubs; in fact very much such a garden, to our non-horticultural eye, as may be seen in any of our American cities. In good weather this place was undoubtedly, as with us, a resting place for the aged and invalids and a lounge for children and their nurses. But when we saw it a raw wind was blowing and the benches were tenanted only by a few old men of a semi-official appearance. The wind did not linger about the place but hurried by to the southward in quest of a better climate, so we followed suit and turned to the right into the space appropriated for live stock.

The cages were covered with wire netting, and were about ten feet square. At the back were little red houses. In the first pen were White pea fowl, some of them very fine, and among them a cock, with red ears and greenish black legs, that bore quite a strong resemblance to a Brown Leghorn. A small B. B. R. Game hen was running about.

In the second pen were a number of White Pheasants of the size common here, and an immense cock bird, of the same variety, that stood head and shoulders above the rest and would weigh as much as any two of them. Running with the pheasants was what looked like a rumpless White Leghorn; a fowl with perfect Plymouth Rock shape and plumage but also rumpless, and a big black rumpless cock with feathered legs, a crest nearly four inches high and a thick beard. Over this coop hung the label "Chine."

In the third cage was a bird of the shape and carriage of a Hamburg. His comb was rose, his hackle a light yellow; the rest of his plumage a light yellowish brown with a bar of darker color on the wings. This fellow had blue legs and half white ear-lobes. The fourth section contained Japanese Bantams, and to the front was attached the label, " Coq et poules blanc, Nangasaki, Japan." These birds were much too heavy for any fancier's idea, and had both red or scarlet and yellow legs, four point combs and black tails.

In the next coop was a huge *pink* fellow like an exaggerated heron. Affixed to the wire of the cage was the legend "Ibis rouges, Para." In color this bird was what we have described him *pink*, not red or any other color, but pink.

Under the title of "Coq et poules de Bentam, Argentes, Java," were some fowls in plumage very much like the S. S. Hamburgs. They were about half the size of the Hamburgs and had red ear-lobes. The cock's breast was white and those of the hens yellowish. In the next was a more interesting display. "Faisans de Mongolie" said the placard. These birds were unlike any pheasant we have seen; they were something like a Ring Neck Golden Pheasant, had long white stripes over the eye like lengthened human eyebrows. The top of the head was dove color and the breast a deep cinnamon or cherry brown, wings cinnamon. Taken altogether they were the handsomest pheasants we had ever seen.

The Australian stock came next and consisted of a pen of Sultans. The cock showed a red face and enormous white earlobes while the hens sported black faces and small white earlobes. Rose combs were seen on both sexes. The Golden Polish, Silver Polish and Mottled Polish came next but were not specially noticeable. In the same pen with a fine collection of California quails was a thing unlike any other creation of the fowl sort we have noticed. In general construction it is a crow but somehow a few bright green feathers show themselves on each side. It looks much as if a painter had used a trifle of window-blind paint to bedevil the bird, or perchance a dyer had tried a new dye on him. How such a queer sight was brought about our very faulty French prevented our ascertaining.

The next coop held some White Holland Turkeys fully up to the *Standard* requirements of the American breeder. Passing downward through the grounds we came upon a "Grue Couronneblette" with a crown of spike-like feathers standing stiffly up from the head, and having intense cardinal colored tail and wing-bars. His height was about three feet, sides white, neck ash color, and his face or ear-lobes white on the upper half and red below.

Passing on, our notes speak of Buff Cochins of the English types, good but small, and Muscovy Ducks, with other specimens of no special note.

At the pond were ducks and swans in great profusion. Among them were the splendidly colored Wood Ducks of our own States and the rare Mandarins, with both of which recent chromos issued by the Poultry World have made our readers familiar. But the sprightliest of all was a variety, about half the size of a Rouen, having a magnificent brown "Bismarck" head, with white tail coverts, and a general cinnamon color throughout. Sleeping on the shore were other ducks of the same general style as the one just described but with whitish yellow heads, and of greater size. In the same inclosure, was a White Crane with jet black legs, and a sort of Guinea fowl we had never previously seen; the neck a bright blue with a brown ruff all round, a clean white stripe on each feather and an azure blue breast. It was described as a "Pintado Vulturine," and credited to Zanzibar.

Of the French class of domestic fowls there was a good showing, one old cock—a Creve Cœur—weighing, as near as we could guess, nine pounds. There was also a showing of Brahmas of the English type, but nothing worth notice.

We wound up our saunter through the grounds by an inspection of the ostrich and emu in adjoining pens. This latter fellow stands about three feet high and has a dirty brown plumage parted down the back like the back hair of a Bowery swell. He is rumpless, and has a short and very heavy triangular black bill.

The Marseilles Zoo is well filled and with a good variety of stock and yet as we turned from the gate to wend our way to the hotel we felt like the boy, there was "lots of news but nothing to holler."

## Genoa.

On the night succeeding our trip to the Zoological Gardens we embarked for Leghorn on the steamer Enna and the next noon entered the harbor of Genoa. On all sides Genoa rises from its semicircular harbor; to say that it lies at the base of a mountain would be incorrect, for it is on the mountain and rises steeply everywhere. Its streets are the perpendicular ways of a nightmare, and run upward till they in many instances "can no more" and halt at the base of a perpendicular rock wall. At various points the grade is so sharp that stairways replace the sidewalks; at others the roofs of houses on opposite sides of the way fairly shut out the light and lack but one or two feet of touching.

Our stay here was too brief to admit of anything but the most cursory look for fowls, but we inspected quite a number that were offered for sale in wicker baskets slung to the shoulders of strapping peasants. As these birds were mostly capons and had the peculiar plumage of their kind, little could be judged from them of the character of the stock we should ultimately find. The next day we again weighed anchor and the following morning were lying in the outer mole at Leghorn. Here we took leave of the courteous Captain Guisippe Graf, whose knowledge of English had made our trip very pleasant, and went through a drizzling rain to the custom house.

## Leghorn.

Our first impression of this city which we had so long desired to reach, was hardly pleasant, but ere the three weeks spent there had passed away, we had learned to tolerate it; not much more however, for it is in winter one of the most comfortless of cities. Imagine a place of nearly one hundred thousand inhabitants, without stoves or any effectual means of heating. And this too when the thermometer is often read at twenty degrees above zero. The houses are of stone with brick or stone floors and in the majority of cases carpetless. Some pampered people indulge in a strip of carpet three feet long and a foot wide at the bed-side, but this is seemingly considered a token of a weak yielding to foreign ideas and is not encouraged. To people of American ideas Leghorn is a good place—to leave.

But our mission was not to growl at the utter absence of sunny skies and balmy winds, but to get our live stock and retreat as soon as possible. Our first acquaintance was with the house of F. Cibo & Co., who handle all the American vessels that come to Leghorn and are proficient in the English tongue that sounds so grateful to the traveler of little French and no Italian worth mentioning. From these gentlemen we received the most material aid, for not only did they furnish us one of their men as interpreter but on numerous occasions the senior owner gave us his time and care in our business transactions. With his assistance we were enabled to describe what we sought to the dealers and obtained better terms than would readily have been accorded.

For the benefit of the uninitiated we remark here that the work of such a house as we have described is comprehensive. When a vessel comes into the harbor one of the firm boards her and pilots her to her moorings. Their boats lie alongside at all hours and will execute any commissions. The provisions of all sorts are ordered through them and no matter what the demand they always find means to fill it, whether a roast is wanted for the galley or a shawl for the captain's wife; or a trip to Pisa is planned. When we reflect that all the time when our trips to the market occupied half a day, these other avocations must be steadily carried out, it is possible to comprehend the extent of the unrewarded politeness extended us on the strength of our nationality and a simple letter of introduction.

These matters we have detailed as a part and parcel of the incidents of our trip and also because they throw light on the conditions under which we worked.

## The Market.

Around Leghorn are built massive stone walls, through which are numerous gates. At each gate is a force of soldiers of the customs department, whose duty it is to see that nothing subject to the consumption tax levied on all food that enters the city, is brought in free. On account of this duty, which applies as much to live fowls as a flask of wine, the market where fowls are collected stands outside the gates. When poultry is shipped from the harbor it must pass in at one gate and out at the opposite one on the harbor side, but by the payment of a small fee a guard can be procured to see the birds outside the city limits, when no tax is levied.

The market is a brick building. In the front are hampers made of round sticks and about four and a half feet long by three wide and one deep. In a hamper of this sort as many as seventy-five or a hundred fowls are sent a three days' journey. The hampers are piled one above another, and the men in charge make nothing of sliding a fresh one on the top of one bristling with heads stretched up through the slats. The hamper goes on with a crash, and the novice looks for broken-necked hens, but he never finds one.

## How The Market Works.

The proprietor of this market explained to us that he had hand-carts out in all directions for the purpose of collecting fowls. Many of these carts do not go more than fifty miles from home, but occasionally they extend their trips as far as the Adriatic on the opposite side of Italy and southward as far as Rome, one hundred miles away. Fowls thus collected are brought to the market, or magazine, and there held to wait the day when the steamer carries them to Marseilles. Steamer days are Tuesdays and Saturdays. On these days a train of drays piled high with crates of poultry are sent under guard through the town. One dray carries about fifteen crates and the line generally comprises from five to six or seven drays.

It will readily be seen from this that the market if visited early in the morning of steamer days gives a pretty full display of all varieties of poultry to be found in a radius of fifty miles from the city. It was our practice to go out on these days and look over all the stock. Whatever we selected was placed one side and reserved for us. In this way we reserved for closer inspection many samples that finally were rejected and let go with the rest to market.

It is a proof of the strong hold of the shape which we call peculiar to the Leghorn breed, that in the thousands of fowls we examined very few differed materially from this style. Occasionally a bird may be seen that is a little more Dorking shaped than the others but it is usually found on examination to be an old hen.

## Farm Poultry.

There is probable no country in which farmers make such a business of raising poultry as in Italy. Back of Leghorn the land is very flat for nearly twenty miles. Through this country run numerous canals fed by the river Arno which connects Florence and Leghorn. The land though nearly as flat as an American salt-hay meadow is about seven feet above the water. From Leghorn a broad, well macadamized road runs northeast to the old town of Pisa. On each side of this road throughout its entire length are vineyards, whose vines in summer cover the ground between the lines of supports at their roots. At about even distances from each other are square massive houses, and about the doorways of each house are from fifty to two hundred fowls. Few people who have not kept poultry in quite large numbers can imagine the appearance of a line of houses each so compassed about.

It was our practice to go with an interpreter outside the walls and carefully look over all the fowls we saw upon our route. A sample excursion was one along this Pisa road, made a few days before our departure. Starting with a cab and Harry, who has piloted many an American on cruises of greater or less length into the country, we began just outside the gate. Quite a promising flock of fowls were scattered about and we were soon in their midst, and the center of a staring crowd of peasants. On our purpose being explained there was first a decided refusal to part with any of their pets. Next came a decision to let some go at a price corresponding to the supposed wealth and infatuation of a man who came in a cab to buy poultry. Finally after much talk and interpreting we left and walked across the way to see the stock of the next man.

Here nearly the same palaver had to be gone through and as we at last were getting into the cab, the first party shouted out that we might have the birds that we had been looking at. This is merely a sample of the routine that must be gone through if one would select the birds he wants from the parties who bred them. With the proprietor of the market, however, no such "dickering" is of use, for the price originally given was the only one we ever could obtain.

## Price of Fowls.

The market price of fowls varies at special seasons of the year for a short time, but in the main is very uniform. So much is this the case that the price quoted to us by parties who had been in Leghorn five years before were the figures when we arrived and during the time of our stay. In capons indeed there is more variation, and at certain celebrations when it is the rule that everyone should eat capon, the prices run to a high figure. At times extra large capons have been sold, we were informed, for as much as four or five dollars and not infrequently eight or ten francs is charged. When we learn that the steady wholesale market rate for common fowls is from two and a half francs for young, to three francs for old birds the value placed on extra nicety is seen.

This idea of charging less for a young and tender pullet than for an old hen is one that might suit a typical boarding house keeper, but seem a little queer to the general reader.

## Food for Fowls.

The universal food for farm poultry seems to be corn. This corn is utterly unlike the large kerneled red corn which is raised on our western farms, and more resembles the white variety of pop-corn common in New England. This was of course the principal food of the stock we brought across, though we supplemented the diet with potato skins, and half a barrel of white meal that had become damaged. It was rather ludicrous to see the surprise of the birds, when on reaching Baltimore, a supply of the large, red Western corn was thrown to them. It did not take them long, however, to decide that red corn was just as good as any other.

The refuse from the house, which we save and give our feathered stock, is in Italy left anywhere it is handiest and the fowls, which have the range of the house as well as the grounds, help themselves. In summer the fowls probably do good in destroying the various insects which attack the vines, but as our visit occurred during the winter we saw nothing of this.

## An Old Story.

Since returning from Italy we have been constantly on the *qui vive* for any information that might lead to the explanation of the fact that Rose Comb Leghorns were among the first imported to Mystic River, Conn. At the time of our trip it seemed that we had definitely ascertained that such was the case. But the "bottom facts" were still a step off. This we learned but a short time ago through Mr. George Burrows who was the possessor of the first lot of Leghorns. His story is substantially as follows:

"I remember well the first lot of Leghorns that were brought to Mystic; there were three of them, a cock and two hens, but on the passage the cock died. Mr. Albert Stark and I were at that time in the 'hen business' and we took the fowls off the vessel lying right down here at St. Jago wharf.

(St. Jago is a tract lying near the river and the wharf was at one time crowded with the shipping that made Mystic famous.) As we had no cock to put with the hens we did not know what to do, but, finally, hearing of a Mr. Baker in New Jersey who had Leghorn fowls—or as they were then called Andalusians—we sent for a cock. When he came we found he had a fine Rose Comb. This feature has cropped out in the original Mystic stock to this day." At this point we tried to ascertain why the old Leghorns were called Andalusians when Andalusia is a part of Spain and a long way from Italy but could elicit nothing but that the fowls since called Leghorns were at first called Andalusians.

## Rose Combs.

From this it will be seen that provided the cock sold by this New Jersey Mr. Barber was from Leghorn—as there is no reason to disbelieve he was—the purity of Rose Comb Leghorn is as fully established as that of the single comb variety; and breeders of Rose Comb stock are as justly entitled to the admission of their fowls into the list of *Standard* birds as any others. This is of course provided the markings are such as would be accepted as those of a Standard Single Comb Brown Leghorn and the only difference is in the comb. Until a like plumage and equally certain reproduction can be attained we cannot, of course claim as high honors for them as for the single comb stock.

That we may rationally ask for a change in the *Standard*,

to include the Rose Combs there is little doubt, though we must acknowledge that in our own breeding and in the experience of others they do not breed quite as true as the single combs. Years of neglect and the killing of Rose Comb birds as culls have naturally put back their development in fine markings or rather stopped them on the progressive scale on which they were being raised. But the genuine material is still there and can in a short time, now the attention of poultry breeders generally has been called to the question, be worked into uniform appearance.

On the matter of the occasional cropping out of Rose Combs we had proof in a bird we own ourselves which is full brother to a single comb Leghorn cock, yet has a heavy Rose Comb and throws about sixty per cent. of Rose Comb chicks.

## American Rose Combs.

That the Rose Comb Leghorn is essentially an American bird there can, from our observation abroad, be little question. As many fanciers are aware, the first impetus of the importation scheme arose from the desire of the writer to obtain imported stock of this variety. Hence when the field of action was at length reached the first inquiries were for Rose Comb birds. These inquiries met with no success. The friend to whom we applied had never heard of any such thing as a double or Rose Comb fowl but the dealers might know of such birds. The dealers were in turn questioned and the result was the same; they had never seen them. Here was a nice situation for one who had traveled so many thousand miles with the vision of splendid Rose Comb Brown and White Leghorns constantly before his eyes. Still we could not give up the idea that somewhere in our search for stock we should stumble on a Rose Comb of some sort, Our ideas had collapsed considerably and we no longer expected splendid plumage, and would have gladly accepted a fowl that could be bred to feather by a couple of years of care. Rose Comb stock certainly had existed in Italy (and presumably in Leghorn or near it) thirty years ago and why not now. So we watched the market and took walks and rides outside of the city, stopping at every house where fowls were kept and inspecting stock, but always with the same results—no Rose Combs.

As may be imagined this dearth of the variety we personally were most interested in caused a great amount of speculation as to why the Rose-Comb had become extinct.

It was after a residence of nearly two weeks in Leghorn that a glimmering of the truth first occurred to us. We were dining with an Italian friend and as a breast of capon was served we noted a sauce containing little dice of pullets' combs. The secret was solved. Single combs are the only ones that would permit of such usage, and as all poultry in Italy is apparently bred for the market, the demand for edible combs is met by the exclusive rearing of single-comb stock.

## Recent Leghorn Importations.

We have recently noticed the advertisements of many parties who claim the possession of Rose-comb stock recently imported from Italy. That such an importation is impossible we should be very sorry to say, for though we have devoted our best energies to the discovery of Rose-combs our result is but a negative one, logically speaking. About Leghorn we feel sure, but of the country on the French borders we have little knowledge, though we examined the poultry of Genoa. It is possible, of course, that in the mountains of the border may be found Rose-comb birds of some sort. As all know, the characteristic of all French breeds is a heavy, a very heavy comb, and just as the inhabitants of any border ground partake of the character of both nations, and speak a mixed language, unintelligible alike to the French from Paris, or the Italian from Leghorn or Rome, breeds may be mixed in form.

This is, however, a supposition, whose truth may never be proved. From our own observation nothing seems more probable than that single-combs are universal throughout Italy.

## Red Leghorns.

From the time that the name Leghorn was first applied to the fowls brought to Mystic, the epithet Red was also used. As a description of the original stock, in the male birds, as well as of the males now to be found in Leghorn it is far more applicable than the word Brown which the change in the fowl has made necessary. In the adult males to be seen in the market at Leghorn, and in the flocks which can be seen at every house on the Pisa road, red is the color. On backs, tail-coverts and wings, it predominates, and is even found in the hackle to a greater extent than in any American strains. On the wing the steely blue bar which is so strongly marked in our best fowls is wanting either in part or wholly.

In the hens something of an approximation to the regular plumage of this variety is to be seen, but the back shows large half circles of lighter color, like the plumage of a Partridge Cochin hen. Were this all, the birds might readily enough be bred into our strains with marked advantage on the score of fertility, but their legs are an insuperable bar to their use in any such way, for they vary from greenish black to lighter willow. It is not by any means an impossibility to get yellow-legged fowls but in getting yellow legs we lose in plumage. The existence of a willow leg on a fowl of this variety will be a matter of great surprise to almost any fancier who has watched the course of breeding Brown Leghorns for some time. A black or willow leg has long been thought proof positive of impure blood; a reversion to some Black Breasted Red Game ancestor, and older fanciers have again and again impressed inquirers with the ruling that willow legs meant mongrel stock, inevitably. To a certain extent this is true; reversion will seldom overleap to the original characteristics after so many years of care. Yet we feel a spice less of certainty in the absolute truth of the law we in common with others have laid down. As the old lady said "seeing is believing" and the willow leg on a "Red" Leghorn hen in Italy, has stared us in the face too often to be readily forgotten.

## White Leghorns.

We have spent much time in wondering how White Leghorns came first to be brought to the United States. Of their first arrival but little is known, or rather but little has been written. That they are the rarest of all fowls in Tuscany is the fact, at present. Even the Dominiques of which we succeeded in finding but six, all told, are as plentiful as the whites; and more so. In the market we never saw a pure white fowl of any sort. Whites with an occasional black flight feather were seen rarely, but a pure white never. On one occasion we certainly thought we had found a *bonanza*, at a house about two miles from Leghorn, for there were two or three white fowls in the flock. We halted immediately and with our interpreter instituted a search, but only to find that the cock—a splendid shaped fellow—had red ear-lobes and some leg feathering; or rather stumps of feathers in his legs. In addition to these "outs" we perceived on close examination that the feathers of the saddle ended in small light yellow spangles. We passed on.

## Black Leghorns.

The black fowl is *par excellence* the favorite fowl in Italy. This can hardly be more of a surprise to our readers than it was to the writer. With the exception of one flock of Blacks, of quite recent importation, there were, as far as we knew, very few fowls of this variety in the hands of American poultry fanciers. Black Leghorns had been made—and very cleverly too—from the sports of Dominique Leghorns. It may not be generally known that Dominique Leghorns throw black sports from time to time, just as the Plymouth Rocks do, but not with any such frequency. These sports if bred together will in time produce a genuine breed of Black Leghorns, true Leghorn in every particular. With these birds, however, we have little to do in considering the Black Leghorn as found in its native haunts.

It may give a better idea of the prevalence of black over all other colors combined, in the birds to be seen by the thousand just outside the gates of Leghorn, if we state that nine out of every ten are jet black without admixture of any other color. It may naturally be asked how if Blacks are so common the first importations were of Browns and Whites. For this we can assign two causes. First the Browns and Whites may have been brought, as rarities as well in Italy as in this country. Secondly that the character of the poultry of a country changes with the varying course of commerce and the consequent demand.

We need but look at the immense disparity between the fowls of America, now common and those of a quarter century back to see this. At the time we speak of, though the Bramah-pootras and Leghorns were known to a few, their influence on the common stock was as yet *nil*.

Time, which has so interweaved the blood of Asiatic and Leghorn with the ordinary barn-yard fowl that you may now cruise the whole day long and never see a single specimen of the old-style bird, has been busy on one side of the world as well as on the other. Though the poultry of Tuscany has been subjected to no such systematic training for specific ends, it must, perforce of altered circumstances, have suffered some modification. Hence to-day, in the section whence came once Brown and White fowls, the Black reigns well nigh unrivaled.

## Earlier Importations of Blacks.

We have visited Mr. Reed Watson at his fine fruit farm of forty acres, at East Windsor Hill, Conn., since our return, and spent a day very enjoyably. To him must be given the credit of being the first to introduce and breed Black Leghorns in this country. He has been a continuous importer and breeder since 1871. In 1876 he got, from Leghorn, by his brother-in-law, Captain Tapley, who is well posted on poultry, a trio, cock and two hens; and in 1878 two cocks and four hens—the last giving the largest per cent. of good chicks.

We had known Mr. Watson for some years and had heard of his noted flock, but we were not prepared to see so fine a display. In anticipation of the change in the Standard to "Black or nearly so" he has bred to that ideal the past season.

Mr. Stoddard speaks, in the Poultry World of May, 1879, as follows:

"Black Leghorns were introduced into this country in the

fall of 1871, by Mr. Reed Watson, of East Windsor Hill, Conn. In 1876 he made another importation more satisfactory in result. In 1878 he got from Italy a cock near to perfection, from which he bred a fine flock—the best he ever had. We visited at his farm in the fall, and congratulated him on his success; and were so much pleased with the high character of his stock that we obtained a drawing, and produced a chromo representing birds from his last importation.

He deserves and should receive commendation for the perseverance and persistence he has manifested in bringing this breed to such perfection. Mr. Watson has sold large numbers of this breed, and has thousands of letters relating to them, in many of which are expressions of satisfaction and praise of his birds. In laying qualities, vigor, hardiness, ease of raising and beauty, they can hardly be excelled, and should be more extensively introduced in the yards of fanciers and others who keep poultry. They are highly valued by those who have bred them continuously."

## Leg Color in Black Leghorns.

The recent convention of the A. P. A. at Indianapolis caused the section relating to the color of the leg in Black Leghorns to read "black or nearly so." When we left this country the *Standard* read "black or yellowish black in front." This does not specify the color of any part of the shank except the front, but gives the impression that yellowish black is the color elsewhere. With the knowledge that a vigorous attempt to change the *Standard* to read "yellow" instead of "yellowish black," would probably be made at Indianapolis, we sought with great care to get yellow legged birds. In this, however, we had little success, though in a few cases we obtained specimens with a flesh colored or light yellow leg. In the majority of cases the yellowish black or black prevailed. That this is the natural color there can be little doubt, but we feel sure that, sooner or later, the fancy which demands a clean yellow leg on many of our black fowls will bring about a change in the Black Leghorns. Still this is probably a thing of several years hence, for some one must first get a flock of yellow-legged birds of this variety. A small number of yellow-legged birds might, with care, be selected in Leghorn, and from them a large number of chicks reared, and a strain with this distinguishing characteristic established.

Mr. Watson to whose flock we have already referred, dissented strongly from the views we have expressed above, and stated that he had tried for years to create a strain of this sort but without success, as yellow-legged birds almost always threw the yellow leg with a touch of white in the plumage. However, we do not give up hopes that some time in the not distant future such a flock can be established.

We are bound, however, in fairness to our readers to say, that the black fowls with clear yellow legs, in Leghorn, were almost all spoiled by a white spangle somewhere in their plumage.

## Non Sitters.

The American fowls are non-sitters. If the original birds were the same and there are no other varieties in Italy how are chicks raised? Turning back to the original stock of 1852, we learned from Mr. Burrows, whom we have already quoted, as follows, substantially: "The first Leghorns I had were bad sitters. Some of them were always wanting to sit but they never would hatch. Always sat about two days and then gave it up. After a few years they gave it up altogether."

This corresponds pretty accurately with the experience of Leghorn breeders, even of late years. Few can show a flock in which there is not a hen which will occasionally cluck and be broody for a day or two, though easily broken up.

This account however hardly corresponds with that of some peasants who pointed out some light yellow hens as the mothers. These hens were a trifle larger than the rest of the flock but had the same general characteristics. Whether they are maintained as a separate flock for the purposes of incubation we did not learn. It is possible, as was suggested by an eminent poultryman with whom we have discussed the subject since our return, that the sitting instinct though faint is sufficient for all purposes while the fowls have perfect liberty among the vineyards, but is lost in the caged life of this country.

It is well known that the Hamburgs are as thoroughly non-sitters as any variety of domestic poultry, and yet, but a few months since an account was published, in the Poultry World, of one which stole a nest and hatched out a fine brood of chicks.

There is little question in our own mind that a flock of any variety of poultry if turned loose on a tract of wild land, would, instead of never increasing and gradually dying out, speedily acquire the incubating instinct. Where man's intervention is removed and the selection for plumage is replaced by the natural selection of the fittest for reproduction a race of sitters and mothers will be speedily produced. A flock of Leghorns turned adrift as we have supposed would never, in our opinion, produce strong sitters; but that they would die out for lack of an instinct which nature has given all wild fowl, in a greater or less degree, we doubt.

This has, indeed, but little bearing on the birds of the fancier which lack the conditions which would foster the sitting instinct. Yet is worth remembering when a pet hen shows, for a day or two, signs of broodiness.

## CAPONS.

Caponizing is fast coming into favor in this country, and is gradually but surely becoming a source of greatly increased revenue to those who rear poultry for market as a business. In France capons are quite common, and bring a high price; but in Italy caponizing is wellnigh universal. It is a matter of great difficulty to obtain adult males, of any variety, that have not been subjected to this process. This is not to be wondered at when we learn that a cockerel is worth, in the market, only from a quarter to a third as much as a capon of the same age. How much this is brought about by the demand of the French market, to which nine tenths of all the fowls raised within one hundred miles of Leghorn go, cannot be known. We fancy, however, that though the French may have set the fashion the Italians have been very ready to adopt it. It is, indeed, simply a question of producing the best article at the least cost. A capon of six or seven months old has consumed probably not a particle more food than the cockerel of the same age, but will dress nearly double as much meat; and of a finer flavor.

In personal appearance the capon has been aptly characterized, by some one as silly looking. He has been dubbed and a crop of silky feathers seems to lie loosely on his hackle; his tail droops and the head is never raised much above the line of the body. Sidling about among the hens in a sort of grown-up chicken-hood, with a voice of querulous old-mannishness, he bears little likeness to the lusty cockerel. There is'nt much activity about the capon, but then neither is there about the prize-pig as compared with the antelope-style porker we have seen in the Southern States.

## Leghorns as Table Fowls.

Though Leghorn breeders in this country have often descanted on the superior qualities of their favorites as table fowls, poultry breeders generally have rather scouted the idea of their excellence in this particular, and while granting them due honors as unrivaled layers, have held heavier fowls, such as Brahmas and Plymouth Rocks, in higher esteem for table purposes. It it rather a striking commentary on this opinion that the French, whom we have always credited with a fine taste in such matters, consume thousands of the Leghorns every week and pay a good price for the privilege. In what way the interior of France is supplied we know not, but can speak for Marseilles which receives from the port of Leghorn alone, two shipments each week. Counting the fowls sent at a very low estimate, it is safe to say that two thousand are sent weekly. Of these about one quarter are capons and the rest pullets from three to four months old.

We have eaten our full share of three-year old hens, gigantic and—breathe it not aloud—tough Brahma cocks and other developments of farmers' poultry, and know pretty well their flavor, but no such specimens ever bid defiance to our teeth on Italian soil. At the little railroad station at Staghno, as well as at the Giappone, in Leghorn, there was alike nothing but tender and delicious poultry. In this, however there is a little secret which may not come amiss to our lady readers who are also housekeepers. As ice is well nigh unknown and fowls cannot be kept for any length of time, all dead poultry is slightly boiled before it is stored in the larder. It will then keep for days and be as tender and fresh as ever. This process we did not see, for we have no "Herald back-stairs reporter" instinct, but learned it from an English lady, for many years a resident of Leghorn.

## Hardiness of Imported Stock.

There is a decided impression in this land of steady habits

that Italy is "sunny Italy" the year round. This idea we shared in common with many others, though perhaps not to so great an extent, as we had made trial of a Florida winter and learned that the land of flowers and and balmy nights existed only in the mind of the enterprising special correspondent. The truth of the matter is that during our stay the thermometer registered as low a temperature as on the New England coast, there were two snows and water was frozen in the gutters. To fowls that are kept mostly at large, this climate is as severe a trial as our bitterest cold to an Americanized stock of the same variety, that are ordinarily well cared for.

We are told that the past winter has been an exceptional one all through Europe and so we can hardly judge of the ordinary conditions that Leghorn stock is subjected to, but it is certain that the birds we inspected showed few traces of suffering. We may safely assume that while there is a decided difference in the climate of Italy and the United States, fowls bred there and imported will suffer no more than the birds of former importations now acclimated here.

Though our own experience is limited to the few weeks that our stock was recuperating before shipment to its various owners, it points strongly the hardiness of imported birds. Of the sixty-three birds which arrived at Mystic, not one showed signs of sickness during all the time we had charge of them. During this time there were two quite heavy snow storms followed by melting weather and the conditions of roup were fully present. Not even the scrawny specimen nickhamed 'Baldy Sours," seemed to feel at all troubled.

This last is a queer specimen, with an eye out and two toes missing, that apparently gave up the ghost about half way across the Atlantic. We found her lying flat on the bottom of the coop with a hen standing on her head. After some trouble we pulled her out and prepared to throw her overboard, had even given half the swing that would have thrown her, when she opened her solitary eye and winked. This is a "petrified fact," to quote the celebrated M. T. We tossed her down on the deck and in less than ten minutes she was fighting through the bars of the coop with some of the other stock.

## Prolificacy.

There can be little doubt that imported stock will lay more eggs in the same time than the descendants of former importations now do. The testimony of all the breeders who received Leghorn stock direct from Italy a number of years ago is the same on this point. Not long since we spoke of this extraordinary prolificacy to a well-known breeder of this variety, and learned that he received some nondescript Leghorns only a few years since, and that most of them layed themselves to death, dying on the nest. Before coming to this unfortunate finis most of them had performed extraordinary feats in the egg-producing line.

We are well convinced from our own study of the subject and what we have seen in our own yards, that if it is safe to expect one hundred and seventy-five eggs per annum from well-bred *Standard* hens, imported birds will for the first two or three years will lay from two hundred to two hundred and fifty. A hen is worth just what she will produce, and if imported fowls will give us the materials for a greater number of thorough-bred chicks, they are proportionately more valuable, to say nothing of the great influence that will run down from generation to generation for a series of years, from the influence of their cockerels on other strains.

## Our Advertisers.

As our readers will perceive, the last few pages contain the advertisements of those who took stock in the enterprise. They are "good. men and true," every one of them, and have proved themselves by many acts ready to lay out money and time freely to gain the very best stock to be had. That the man who is ready, at any time, to pay a high figure to secure extra birds is the man to buy your good birds from, is our experience. One of our advertisers told us but a few days ago that he was always ready to pay a high figure, which he named, for a Brown Leghorn cockerel that would honestly score ninety-six points. He has had many birds sent on approval and he has seldom retained one, simply because he could not honestly score the specimen up to the required point. This is a particular man in buying but he is just as careful of what he sends out to his customers.

This is but a sample of the men whose addresses may be found farther on and we feel confident that those who deal with any of them will be pleased at the honest and gentlemanly treatment they receive.

# STOCKHOLDERS.

Those marked * will not breed the imported Black Leghorns.

A. B. COGGRSHALL, Newport, R. I.
THEO. KENNEY, Scottdale, Pa. *
MORGAN PIERSON, Clinton, Conn.
GEO. H. FANCHER, Winsted, Conn.
CHAS. S. HASTINGS, St. Johnsbury, Vt. (2 shares.)
A. B. CAMPBELL, Norwich, Conn. *
J. E. CLAYTON, Saginaw City, Mich.
GEO. H. TOWLE, Truxton, N. Y. (Dominiques.)
W. B. EVANS, Ripley, Ohio.
W. N. CROFFUT, Binghamton, N. Y.
WM. E. HART, East Cleveland, Ohio.
C. R. HARKER. Rochester, N. H. *
W. R. WILSON, Whippany, N. J.
E. B. TOWLE, Newburyport, Mass. *
S. J. FEARING, 91 Commercial Street, Boston, Mass.
P. H. MARLAY, Lincoln, Nebraska. *
T. P. SNYDER, North Adams, Mass. *
WM. BOWER, New Britain, Conn.
J. M. CLARK, Box 447, Amherst, Mass.
D. ANDREWS, San Jose, Cal.
GEO. M. SNOW, Box 1076, Salt Lake City, Utah. *
H. L. FISK, Box 859, Worcester, Mass. (1½ shares.)

☞ In sending letters of inquiry to any of these gentlemen please enclose a stamp. Remember the cost of paper, particularly when Uncle Sam has put his mark on it.

*Motto: Luke vi. 31.*

# Golden Rule Poultry Yards,

CHARLES S. HASTINGS,

Proprietor,

ST. JOHNSBURY, - - VT.

---

VARIETIES:

IMPORTED BLACK LEGHORNS,

SILVER GRAY & COLORED DORKINGS,

LIGHT BRAHMAS,

BROWN LEGHORNS, "BLACK DIAMONDS,"

AND

"GOLDEN-WINGED BELLFOURS."

---

**EGGS AND FOWLS IN SEASON.**

*I make it a Point to live up to my Motto.*

Descriptive Circular and Price-List sent to any Address.

[SEE NEXT PAGE.]

*Motto: Luke vi. 31.*

# Golden Rule Poultry Yards.

*My Yards of Rare and High-Class Poultry for 1880, are Mated as follows:*

### Imported Black Leghorns.

Cockerel from Reed Watson, from his importation of 1878, mated with 6 Hens imported personally by Mr. F. H. Ayres, direct from Italy in 1880. I believe this yard is not excelled in America. EGGS, $2.50 per 13, or $4.00 for 26. CHICKS in the fall, at $15.00 per trio

### Silver Gray Dorkings.

Cockerel from my stock, selected from some thirty raised last season, and scored by H. S. Ball at 92 points; is a large and very fine plumaged bird, mated with four hens from Warren, Conn., four nice pullets from Bowen, Mass., and two pullets raised by myself and scored by Mr. Ball at 92 and 93 points. EGGS, $2.00 per 13, or $3.50 for 26.

### Colored Dorkings.

Cockerel from Bowen's stock, mated with two pullets, Blakeslee stock, and eight nice pullets of my own raising. All large and evenly marked. EGGS, $2.00 per 13. $3.50 for 26.

NOTE.—Mr. H. S. Ball, A. P. A. judge, gave me the credit of having the best flock of Dorkings he had seen together at any show this winter.

### Light Brahmas.

Cock, Josselyn's strain, mated with seven pullets, part Josselyn's and part Burnham's stock. All very fine birds.

Cockerel, Josselyn and Burnham, mated with three hens, Burnham, and three hens, Josselyn. "You pays your money and you takes your choice." EGGS, $2.00 for 13, or $3.50 for 26.

### Brown Leghorns.

A friend has a very fine flock of these "egg machines," that I will sell eggs from at $1.50 per setting. Cockerel, Banks' stock, mated with hens and pullets, Harker's strain. Nicely marked and good points throughout.

### Black Diamonds.

Cockerel, 8½ pounds, mated with two hens weighing 7½ pounds each, and six nice pullets weighing from 6 to 7 pounds each. All nicely marked, and not related in the least to cockerel. EGGS, $3.00 per 13, or $5.00 for 26.

### Golden-Winged Bellfours.

NO EGGS FOR SALE. For a better description of these last two new breeds, established by careful selection and breeding, myself, as well as a general description of the Standard Varieties bred by me, see my new circular, sent to any address.

## CHARLES S. HASTINGS,
St. Johnsbury, Vt.

# LEGHORN FOWLS

## AND

# SHEPHERD DOGS

## IMPORTED STOCK.

## H. L. FISK & SON,

(P. O. Box 859.)

## WORCESTER, - - MASS.

# W. N. CROFFUT,
## BINGHAMTON, - - N. Y.

BREEDER OF STANDARD

# Brown and Black Leghorns.

My strain of **Brown Leghorns** are not surpassed in America for breeding fine combs, solid white ear-lobes, with elegant plumage and symmetry. My Breeding Stock this season is better than ever, and mated to produce the best possible results, and is composed of 4 yards.

My **Black Leghorns** are also very fine. I have some that Mr. F. H. AYRES imported for me. They are beautiful in form and feather, and are admired by all who see them. I am now prepared to book orders for eggs from either of the above varieties at

## $2.00 per 13, or $3.00 per 26.

packed to go any distance, and warranted to hatch a fair per centage. Fowls and Chicks For Sale at all seasons of the year, at Moderate Prices. Write for what you want.

## W. N. CROFFUT,
BINGHAMTON, N. Y.

# BLACK LEGHORNS,
# BROWN LEGHORNS,
### AND
# ROSE-COMB BROWN LEGHORNS.

My Black Leghorns are imported fowls. My Single-Combed Brown Leghorns are from Kinney's Brown Prince, 3rd, winner of the Centennial prize.

My Rose-Comb Brown Leghorns were obtained direct from the originator (McDaniel). They combine all the good qualities of the single-comb varieties, but having double combs there is no danger of their combs freezing, which makes them one of the most desirable birds known.

### *FOWLS FOR SALE AFTER SEPT. 20.*

## EGGS,

| | | |
|---|---|---|
| BLACK LEGHORN | (per 12) | $3 00 |
| SINGLE-COMBED BROWN LEGHORN | " | 1 00 |
| ROSE-COMBED BROWN LEGHORN, | " | 3 00 |

Eggs Packed and Delivered at Express Office in good condition, Free of Charge. SATISFACTION GUARANTEED.

# JOHN W. CLARK,

P. O. Box 447.     AMHERST, MASS.

# MORGAN PIERSON,
## *CLINTON, - CONN.*

IMPORTER AND BREEDER OF PURE BRED FOWLS.

# PLYMOUTH ROCKS,
*World's Excelsior Strain.*

# BLACK LEGHORNS.
*Disaster Strain*, Imported by me in March 1880, and are The Fowl of Italy, and the Greatest Layers in America.

# BROWN LEGHORNS,
*Prince of Wales, and Clift's.* These two strains cannot be beat.

# DARK BRAHMAS,
*Black Prince Strain.*

# HOUDANS,
*Grant and Pinckney Strains.*

# COLORED MUSCOVY DUCKS.

*EGGS AND FOWLS IN SEASON.*

WRITE FOR WHAT YOU WANT.

**Morgan Pierson, - - Breeder.**

# A. B CAMPBELL,
## NORWICH, - - CONN.

### BREEDER OF

# BROWN LEGHORNS

#### EXCLUSIVELY.

ALWAYS RELIAALE.

I have made a specialty of this breed for seven years and to day have as fine a breeding strain as can be found in this country.

## EGGS AND FOWLS IN SEASON.

### DESCRIPTIVE CIRCULAR AND PRICE-LIST FREE.

## SATISFACTION GUARANTEED.

# WHIPPANY STOCK FARM!

I have some GOOD STOCK

PLYMOUTH ROCKS,

BROWN LEGHORNS,

LIGHT BRAHMAS,

BUFF AND PARTRIDGE COCHINS,

**GAMES that are GAMES,**

AND

## Imported Black Leghorns!

*All First-Class Stock.*

EGGS, $2.00 for 13; $3.50 for 26.

## W. R. WILSON,

*Whippany, Morris Co., N. J.*

# GEO. E. FANCHER,

### BREEDER OF

## IMPORTED BLACK LEGHORNS.

—o—

## Rose-Comb Brown Leghorns,

AND

## Rose-Comb White Leghorns.

—o—

The flock of

# BOLTON GRAYS,

of the late Wm. Higgins.

—o—

CHICKS from all these varieties will be ready in the early fall of 1880, and as I have fine breeding stock I can safely promise

## NICE CHICKS.

Write me.

### GEO. E. FANCHER,

## WINSTED, - - CONN.

1880.

# Hoosac Tunnel Poultry Yards.

BROWN LEGHORNS,
BLACK HAMBURGS,
BLACK LEGHORNS,
BRONZE TURKEYS,

**FOWLS FOR SALE.**

DOMINIQUE LEGHORNS,
PLYMOUTH ROCKS,
AYLESBURY DUCKS.

**EGGS IN THEIR SEASON.**

T. P. SNYDER, Proprietor,

NORTH ADAMS, - - MASS.

# J. E. CLAYTON,
## SAGINAW CITY, - MICH.

---

### HEADQUARTERS FOR THE NORTH-WEST!

---

# BROWN LEGHORN
## POULTRY YARDS.

---

I make a specialty of breeding this Beautiful, Profitable and ever Popular Breed. My Breeding Pens are made up of *Standard* Birds, scoring from 90 1-4 to 93 points. I feel safe in guaranteeing perfect satisfaction to purchasers of eggs, which I ship in their season, securely and safely packed, at

### *$2.00 FOR 13, $3,00 FOR 26.*

Orders solicited, booked and sent in turn. No Stock for sale now.

## CHICKS READY TO SHIP BY SEPTEMBER 1, 1880.

Address

## J. E. CLAYTON,
*SAGINAW CITY, MICH.*

# EGGS

## FROM

## WHITE LEGHORNS!

### $2.00 PER SITTING.

These fowls are J. Boardman Smith, and W. H. Todd strains and have been selected and mated with care, to produce the best results.

All eggs sent out are safely packed and Warranted Fresh and Pure.

### CHICKS IN THE FALL.

Write.   Address

## P. H. MARLAY,

*Lincoln,*   —   —   *Nebraska.*

**BROWN LEGHORNS.** { **CALIFORNIA.** } **BLACK LEGHORNS.**

I HAVE THE ONLY

# IMPORTED BLACK LEGHORNS

in the State. I will, this season, sell a limited number of

*EGGS, AT $5.00 PER SITTING.*

CHICKENS FOR SALE IN THE FALL.   ORDER IN TIME,

---

# MY BROWN LEGHORNS,

at present, consists of One Pen of Finely-penciled Pullets, mated with a Deep-colored, Finely-marked Cock that is the admiration of all. Their eggs have proved to be fertile and chicks hardy. The pullets have proved themselves remarkable layers and I shall continue to improve them by breeding only from the best. I shall increase the number of pens next season. No two breeds allowed to run together at any season. Parties ordering can rely upon stock that will prove true to name.

# D. ANDREWS,

*Cor. 11th and San Carlos Sts.,*

SAN JOSE, CAL.

# WHITE COCHINS,

George 1st, (5269) and Western Belle, (5270) with five other Fine Exhibition Hens.

### *EGGS $3.00 PER 13, $5.00 PER 26.*

## PARTRIDGE COCHINS,

Hector, (5271) and Maud, (5272) with five fine hens from the best stock in the United States.

### *EGGS $3.00 PER 13, $5.00 PER 26.*

I exhibited the leading stock of the above pens at the Cincinnati Poultry Exhibition, January, 1880, receiving 2d and highest, and special on Geo. 1st, the other three taking three of the remaining regular prizes.

## OTHER VARIETIES.

I can also furnish Eggs from Houdan (Butler's) and B. B. R. Game stock, at $2.00 per 13. Address

## W. B. EVANS,
### RIPLEY, OHIO.

# MR. HART'S CARD.

Since the poultry interests of America have become so extensive and important a business, Yankee enterprise has crossed oceans and brought to these shores nearly every known variety. It is now easy for each farmer or fancier to procure such varieties as are best suited to his wants, whether it be a large amount of meat, a generous supply of eggs, or the gratification of the eye. Nearly every variety has some specially good quality, while none can claim perfection in all. For twelve years I have bred high-class poultry, and during that time have had nearly every variety of Standard fowls, and have found none equal to Leghorns in the production of eggs. Of this breed, the Browns and Blacks are most desirable in most localities and with most breeders, because less liable to to show soiled plumage. Then too, I have found the Browns more hardy and prolific. Am now breeding from the best specimens of this color that can be procured. My success in growing a large per cent of well-marked stylish birds has been very gratifying. This is especially true as regards cockerels. That they are appreciated by fanciers is evidenced by the large number sold to the best judges, for the improvement of their flocks the coming season.

I am breeding from two yards. One contains a cock and ten hens, the other a cockerel and ten pullets. All very fine birds—real beuties, and so nearly alike as to make it difficult to distinguish one from the other. Grace in form and grace in every motion, a delight to the eye as well as a sure source of supply to the egg basket. Pleasure and profit are the sure result of well-selected stock of this variety.

Rose-comb Brown Leghorns are not so well known. Some say they were originally imported from Italy. Others claim that they are "sports" from single comb birds. Others again, say they are a cross with Hamburgs. I do not attempt any light on the subject of their origin but cannot fail to see that they do remedy the greatest defect that can be urged against the single comb birds, in having a low, rose-comb. Thus exemption from frost is greatly in their favor, and must make them deservedly popular in Northern States. This is no doubt the reason of their being such remarkable winter layer. Six pullets of this variety have given me an average of four eggs per day all winter. They have real merit and must become prime favorites when better known.

A few sittings of EGGS FROM THESE BIRDS,

## $3.00 PER 13.
## WM. E. HART.

*EAST CLEVELAND, OHIO.*

www.ingramcontent.com/pod-product-compliance
Lightning Source LLC
Chambersburg PA
CBHW060004230526
45472CB00008B/1940